Jean Carlo Rauschkolb
Renan Bruno Dal Cero
Priscila Mayara Wust

Evaluation of the Efficiency of Ozone on the Shelf Life of Strawberries

Jean Carlo Rauschkolb
Renan Bruno Dal Cero
Priscila Mayara Wust

Evaluation of the Efficiency of Ozone on the Shelf Life of Strawberries

Sanitisation, Strawberry, Shelf Life, Ozone

ScienciaScripts

SUMMARY

The strawberry is considered attractive due to its sensory characteristics and, above all, its nutritional composition. Its cultivation is of commercial importance, since it can be a fruit that is consumed fresh and is also widely accepted in the preparation of other foods such as cakes and sweets. Due to its lack of skin, its consumption is restricted to the removal of the stalk, which can in some cases subject the consumer to the ingestion of undesirable components, as well as being a product with a limited shelf life. Sanitisation with aqueous ozone, chlorination and the use of ultraviolet radiation are technologies that can be used in the minimum processing of strawberries. The aim of this study was to adapt strawberry minimum processing technology in order to maintain the microbiological and physico-chemical quality of the product over a 10-day storage period. An ozone generator capable of meeting the gas demand required for the experiment was manufactured. The effectiveness of sanitisation methods in reducing the initial contamination of strawberries by filamentous fungi and yeasts was compared. Physico-chemical quality parameters such as pH, mass loss, colour, acidity and soluble solids were assessed. Strawberries of the 'Oso grande' cultivar were selected and the fruit that did not have % red flesh was discarded. The fruit was washed in drinking water, followed by immersion in water, with the following treatments: treatment 1: containing 150 ppm of active chlorine; treatment 2: containing 107 ppm ozone; treatment 3, containing 214 ppm ozone and treatment 4, with ultraviolet radiation only. They were then dried and packed in plastic jars with lids, containing between 40 and 70 grams of sample and refrigerated at 5°C for 10 days. The data obtained was subjected to analysis of variance at 5% probability, with the significance of the means compared using the Dunnett test. The values for the physicochemical analyses varied over time as expected, but there was no significant difference between the treatments applied and the control sample. The microbiological analyses showed better results with the use of ozone than with the other treatments.

Keywords: Sanitisation, Strawberry, Shelf Life, Ozone.

CHAPTER 1

INTRODUCTION

Strawberries are considered attractive because of their sensory characteristics and, above all, because of their nutritional composition. Strawberries contain fibre, carbohydrates, proteins and sugars. They also contain the main minerals, B vitamins, folic acid and vitamin C, as well as being a rich source of phytochemical compounds, represented by polyphenols, which together act as antioxidants (GIAMPIERI et al., 2012).

However, fruit and vegetables carry bacteria and mould directly from the field, which reduces shelf life and can cause serious health problems. Although almost all food processing facilities wash produce with clean water, this alone does not guarantee adequate sanitisation against fungi and bacteria (OZONE SOLUTIONS, 2014).

Therefore, sanitisation prior to sale is a growing processing trend that offers consumers convenience, quality, nutrition and safety. As it involves eliminating or reducing natural barriers to spoilage, the fact of processing justifies it by offering food professionals a huge challenge in trying to extend the shelf life of fresh produce (RAHMAN, 2007).

The main control factors for food preservation, apart from reducing the original microbiota through heat treatment or the use of chemical agents, are the atmosphere, temperature, pH value and the amount of free water available in the food. In the case of vegetables, the use of chemical agents such as sodium hypochlorite, chlorine dioxide, peracetic acid or ozone is promoted (CENSI, 2011).

Chlorinated compounds are among the most widely used sanitisers in the food industry. However, the reduction in microbiological efficiency combined with the potential toxicity of chlorination by-products has made this process less and less attractive (SILVA et al., 2011).

Another means of sanitisation is non-ionising ultraviolet rays, which have been used extensively for many years by industries to disinfect equipment, glassware and the air, as they are a clean source and leave no residue on the disinfected materials (EMBRAPA, 2006). Although it is expensive to install, this prevents it from being used by producers and retailers. Like UV radiation, ozone does not leave chemical residues, nor does it degrade molecular oxygen through reaction or natural degradation. This advantage makes ozone a useful tool for food technology processes. Since ozone is a safe medium, it has been generally recognised as safe (GRAS) in the United States (FDA, 1995).

1.1   OBJECTIVES

1.1.1 General Objective

To evaluate the efficiency of the sanitisation process with aqueous ozone on the shelf life of strawberries (Fragraria ananassa Loran. cv Oso grande).

1.1.2 Specific objectives

- To develop equipment capable of producing ozone gas in adequate quantities for the strawberry sanitisation experiment;
- Applying different strawberry sanitisation methods, including aqueous ozone;
- Carry out physico-chemical and microbiological analyses of the fruit during a period of refrigerated storage;
- To compare the efficiency of the different sanitisation methods applied to strawberries.

CHAPTER 2

LITERATURE REVIEW

In modern society, saving time and the need to incorporate healthy eating habits have become fundamental and are the main reasons for the growing consumption of vegetables, including those known as minimally processed (GERMANO, 2008). In view of this, the commercialisation of most fresh vegetables can be extended by storing them immediately in suitable conditions that allow a reduction in normal metabolism without altering the physiology of the product (CHITARRA, 1990).

In the case of strawberries, long-term preservation with properties similar to those of fresh fruit is still a technological challenge to be overcome. No economically viable method preserves the quality of fresh fruit, resulting in the loss of its peculiar characteristics of texture, aroma, colour and flavour (VENDRUSCULO & VENDRUSCULO 2005).

Hygiene and sanitisation in the food industry are two of the most important prerequisites for food safety. Loss of quality and microbiological changes in food are generally attributed to poor or inadequate hygiene and sanitisation (MORETTI, 2007).

Heat treatments are conventionally used to achieve safety standards. However, undesirable sensory and nutritional changes, such as colour degradation, softening of tissues, loss of vitamins and disappearance of natural freshness, have turned the focus of research to less aggressive processing technologies that extend shelf life without the detrimental effects of severe heating (GUZEL-SEYDIM et al., 2004; KHADRE et al., 2001).

Technologies such as UV radiation and high-pressure electrical impulses have emerged as attractive alternatives to conventional thermal processes.

The use of ozone as a sanitising agent is one of the emerging and challenging technologies with potential application in the food industry (MADAIL, 2008).

2.1 STRAWBERRY CONSUMPTION

Strawberries are considered attractive because of their sensory characteristics and, above all, because of their nutritional composition. Strawberries contain fibre, carbohydrates, proteins and sugars. They also contain the main minerals, B vitamins, folic acid and vitamin C, as well as being a rich source of phytochemical compounds, represented by polyphenols, which together act as antioxidants (GIAMPIERI et al., 2012). From a nutritional point of view, 100g of strawberries provide 39 calories. The fruit contains 90 per cent water, 8.5 per cent carbohydrates, 1 per cent protein, various minerals such as sodium, potassium, calcium, silicon, iron, phosphorus, chlorine, magnesium, sulphur and some vitamins A, C and B-complex (GONSALVES, 1992).

Looking at the specifics of consumption, the United States is the biggest consumer of fresh strawberries, with a volume of 795,000 tonnes, followed by China 452,000 tonnes, Japan 191,200 tonnes, Canada 95,300 tonnes and Mexico 49,500 tonnes. In South America, the best-placed country in the 2008 FAO ranking is Chile, in 22nd place[a] . Brazil is only 54[a] , ahead of Bolivia in 60th place and Uruguay, which isn't even in the ranking. The Latin American countries ahead of Brazil are Colombia in 25th, Peru in 29th, Argentina in 38th and Paraguay in 53rd place (SPECHT & BLUME, 2008).

In absolute numbers, world strawberry production has been growing in recent years. Between 1997 and 2006, production grew by 29%, while the area planted grew by 18% (SPECHT & BLUME, 2009). In 2006, world production was estimated at 3,908,975 tonnes, for a total planted area of 262,165 hectares (FAO, 2008).

The increase in the consumption of minimally processed fruit and vegetables has been attributed to the health benefits they provide, the fact that they retain characteristics close to the fresh state and also to market trends in relation to the

Product	Temperature		
	0°C	10°C	20°C
	KJ/Kg.h		
Strawberry	0,132 - 0,198	0,539 - 1,045	1,012 - 2,156

Source: Adapted from (HARDENBURG et al., 1990).

Temperature is the way to measure heat. It has a very characteristic effect on fruit respiration. When heat rises, respiration accelerates, causing the dry matter of the product to wear away and consequently degrading fruit quality (CORTEZ et al., 2002). Rapid cooling or pre-cooling therefore consists of immediately removing the heat that the fruit brings in from the field, before it reaches its final storage temperature (CANTILLANO, 2005).

Fruit and vegetable packaging needs to be adaptable to the temperatures required by the products. They must remain in contact with the environment in order to acquire the right temperature for their preservation (CHITARRA, 1990). Rapid cooling or pre-cooling consists of immediately removing the heat that the fruit brings from the field before it reaches its final preservation temperature (CANTILLANO, 2005).

2.2.1 Post-harvest losses

In addition to the losses caused by metabolic changes, inadequate handling during harvest and all the post-harvest stages can predispose the product to infections caused by quiescent or opportunistic micro-organisms that can develop in areas that have suffered mechanical damage (CENCI, FREIRE JUNIOR & SOARES, 1997).

In order to maximise shelf life and reduce post-harvest losses, keeping the fruit preserved for longer for consumption, it is important to know and use appropriate handling practices during the harvest, post-harvest, storage, transport, distribution, marketing and consumption phases (FREITAS-SILVA & VENÂNCIO, 2010).

Strawberries are one of the few fruits that are picked, sorted, graded and packed by the same person in the field. This avoids excessive handling, which would cause physical damage to the product, leaving the fruit susceptible to rot (CANTILLANO, 2005).

The different stages of handling and post-harvest operations make a decisive contribution to the useful life of fruit, vegetables and flowers. On their own, they can be responsible for the commercial devaluation of the product by compromising its appearance, aroma and flavour (EMBRAPA, 2006).

Post-harvest losses of all types of food are generally considered to be higher in less developed countries, with tropical areas falling into this category. Vegetable products comprise approximately 25 per cent of the main food crops produced in these countries, including roots, bulbs, fruit and vegetables (CHITARRA, 1990).

Strawberry harvesting is one of the most delicate and important operations in the entire crop cycle. Strawberry fruits are very delicate and not very resistant due to their thin skin, high water content and high metabolism, which requires a lot of care during harvesting. If they are picked too ripe, they may arrive on the market decayed and rotten; if they are picked unripe, they will be highly acidic, astringent and lack flavour. In both cases, the product reaches the market with low commercial value (CANTILLANO, 2005).

They are harvested by hand at the "ripe" point for industrial purposes. Colour is the most important parameter for defining the harvesting point for strawberries. Generally speaking, the fruit should have at least 50% to 75% of its surface bright red when intended for fresh consumption (CANTILLANO, 2005).

Selection is a stage of separating the raw material mainly by its quality, according to the purpose of each industry. The presence of defects (mainly in terms of colour) or the degree of deterioration can be assessed. In general, products that do not have the characteristics compatible with the established standard are excluded (CARDOSO & SILVA,1999).

Food, in general, has to be packaged at some stage before it can be consumed (GERMANO, 2008). Packaging must therefore fulfil three basic functions: to pack

2.3 SANITISATION METHOD

According to Lovatel et al. (2004), the successful production of food products will depend on the hygienic management of the small-scale production system, particularly artisanal production. In this system, none of the stages of the process can neglect the basic principles governing hygiene and industrial health, as well as their technical application, eliminating any possibility of contamination.

Chitarra (2005) states that freshly harvested vegetables must be free of infectious agents, insects, synthetic chemicals (pesticide residues) and dirt before packaging and subsequent marketing.

The main objective of cleaning the raw material is to eliminate impurities and contaminants that are a health hazard or aesthetically undesirable, adhered organic and mineral residues, including dirt, dust, plant debris, controlling the chemical and biochemical reactions that hinder the subsequent process and jeopardise the quality of the final product (CARDOSO & SILVA,1999). Washing and immersion in water not only removes visible dirt but also significantly reduces the microbiological load of these products (ANDRADE, 2008).

When used properly with water of adequate quality, chemical antimicrobial agents help to minimise the potential for microbial contamination of processing water and subsequent cross-contamination of the product (FDA, 2008).

Washing fresh fruit and vegetables before cutting is important to control microbial loads such as mesophilic microflora, lactic acid bacteria, coliforms, faecal coliforms, yeasts, moulds and pectinolytic microflora. Minimally processed products are generally washed at 50-200 ppm

of chlorine or 5 ppm of chlorine dioxide, which can also help reduce browning reactions (PERERA, 2007).

The main control factors for food preservation, apart from reducing the original microbiota through heat treatment or the use of chemical agents, are the atmosphere, temperature, pH value and the amount of free water available in the food. In the case of vegetables, the use of chemical agents such as sodium hypochlorite, chlorine dioxide, peracetic acid or ozone is promoted (CENSI, 2011).

2.3.1 Clóro

One of the most common sanitising agents known is chlorine. Since the first recorded use of chlorine as a disinfectant agent for water in 1896 (Austro-Hungarian naval base of Pola, in the Adriatic Sea), the use of chlorine has spread throughout the world and today 90% of water treatment plants use it. It is a traditional monopoly that no other product has ever achieved or equalled (HESPANHOL et al., 1989).

Its power to destroy or inactivate the vast majority of pathogenic organisms capable of producing disease or other undesirable organisms is well known. Despite the benefits of disinfection, chlorine and other oxidising compounds can react with natural organic matter (NOM) from surface water sources, forming disinfection by-products that can be harmful to human health, such as trihalomethanes (THM), haloacetic acids (HAA) and haloacetonitriles (HAN), among others (HONG et al., 2007).

Although water disinfection can be carried out by various means (heat, ultraviolet light, ozone, chlorine dioxide), the use of chlorine and its derivatives has been the most widespread due to its ease of application, lower economic cost and high efficiency (HESPANHOL et al., 1989), making it one of the most widely used sanitisers in the food industry for hygiene purposes. However, the reduction in microbiological efficiency combined with the potential toxicity of chlorination by-products has made this process less and less attractive (SILVA et al., 2011).

One of the major concerns of the health authorities is that, in an attempt to reduce the concentration of trihalomethanes (THMs) attributed to the practice of chlorination, disinfection techniques will be adopted that offer less security against possible water contamination. Studies on the use of other disinfectants have used ozone, chlorine dioxide and chloramines, which do not produce THMs products (SUN, 2014).

Ozone, as a powerful oxidising agent, is used in various treatments including water treatment, disinfection of equipment, as well as for the preservation of perishable items such as fruit, vegetables and meat (SEYDIM et al., 2004). Table 2 shows the oxidation potential of various sanitising agents.

Table 2. Oxidising agents and their oxidation potentials.

Oxidising Agent	Oxidation Potential (V)
Fluorine (F_2)	3,03
Hydroxyl radical (OH^*)	2,80
Ozone (O_3)	**2,07**
Hydrogen Peroxide (H_2O_2)	1,78
Chlorine dioxide (ClO_2)	1,50
Hypochlorite (NaClO)	1,49
Chlorine (Cl_2)	1,36

Source: Adapted (GUZEL-SEYDIM, GREENE & SEYDIM, 2004).

According to Leme (1990). Some of the advantages of using ozone are listed below:

• Its disinfectant action is effective over a wide temperature range;

• Its sporicidal bactericidal action is faster and greater than all other known agents. It is said to be 300 to 3,000 times greater and faster than chlorine and only requires short periods of contact;

• No odours are generated or intensified because no addition or substitution complexes are formed;

• It can be used to remove pesticides or other organic substances, such as synthetic detergents, herbicides, etc;

• When it decomposes in water, it only produces oxygen;

• Its oxidising power is not affected by the pH of the water.

The main disadvantages of using ozone are basically the high cost of installing the equipment and the use of electricity during its operation.

Unlike other sanitisers, ozone does not leave chemical residues or degrade molecular oxygen through reaction or natural degradation. This advantage makes ozone a useful tool for food technology processes. In the United States, ozone is generally recognised as safe (GRAS), meaning that it is safe to use (FDA, 1995).

Many companies have used ozone as a combined process with other agents. Due to its high volatility, the "ozone bath" disinfects at the same time as rinsing the plant of the disinfectant used in the previous step, without leaving any residue on its surface (OZONE SOLUTIONS, 2014).

In addition, immersing the fruit and vegetables in ozonised water (2 ppm) maintained by microbubbles can reduce the amount of pesticides in strawberries. The removal of phenitrotione residues increased with increasing treatment time and ozone concentration through continuous bubbling. There was a 55% reduction in residues in 5 min and 58% after 10 min. In cherry tomatoes and strawberries, after 10 min of treatment the residues were reduced by 35% and 25% respectively (IKEURA et al., 2011). Table 3 shows the different characteristics of the chlorination and ozonation process.

Table 3.Comparison of the characteristics of chlorination and ozonation processes (- None; + Low; + + Medium; + + + High).

Features	Cl_2	ClO_2	O_3
Security	+	+	+ +
Removal of bacteria	+	+	+ + +
Virus removal	+	+ +	+ + +
Removal of protozoa	-	+	+ +
Toxic residue	+ + +	+ +	-
By-products	+ + +	+ +	+
Operating costs	+	+	+ +
Investment costs	+ +	+ +	+ + +
Formation of toxic compounds	+ + +	+	+
Odour and aftertaste	+ + +	+ +	+

Source: Adapted from SILVA et al., 2011.

Ozonated product has a longer shelf life and less spoilage occurs during transport and on the customer's shelf. In fact, many companies can achieve a 20 to 50 per cent reduction in spoilage, which results in a very quick payback, higher profits, and a significant reduction in the possibility of a public health problem (OZONE SOLUTIONS, 2014). Figure 2 shows the effect of ozone on bacteria, in which the

-NE 555 timer   ;

-IRF3205 transistor   ;

-Thin copper wire   to generate the greatest corona effect;

-2mm acrylic sheets   ;

-Silicone   rubber sealant   ;

-Potentiometer   10 K

-Transparent empolythene   outlet hose   ;

-Diode bridge   ;

-Cables   for power supply;

-Resistors   ;

-Energy dissipater   ;

The device was assembled based on the structure in Figure 4.

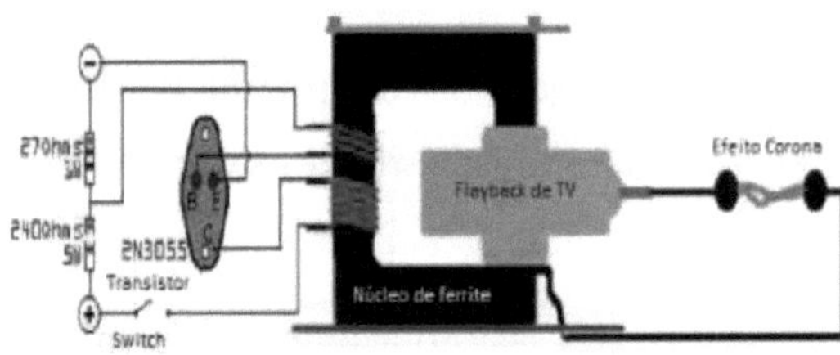

Figure 4 - Ozoniser assembly diagram
Source: Adapted from Wikihow, (2002).

To build the device, the power supply was first connected to the flayback, transistors and resistors as shown in Figure 5.

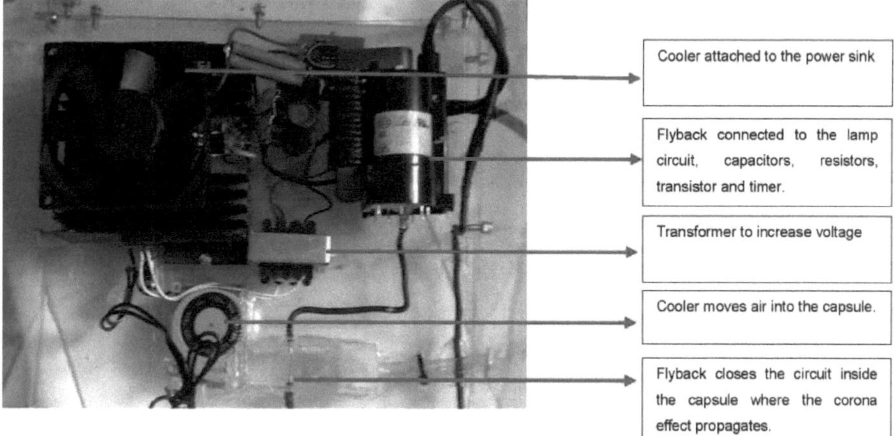

Figure 5 - Ozoniser assembly diagram.
Source: Author's elaboration, 2015

The two smaller sides at the end were made from thin acrylic sheet, where two holes were drilled for the air inlet and outlet, and two more holes were drilled in the sides of the capsule so that the electrical discharge wires could be attached. Silicone glue was used to attach the wires to the hose and the cooler nozzle.

So the cooler was connected to a potentiometer, which is a resistance that can be adjusted by means of a cursor, which is in contact with a resistor connected to two terminals. The moving contact of the potentiometer moves from zero (or minimum resistance) to the maximum value, adjusting the speed of the cooler and consequently increasing the air flow into the capsule (Figure 5).

To make the set work, the potentiometer was switched on so that the air moved through the capsule. In order to also keep it cool, the cables were connected to the power supply, then the amperage of the supply was adjusted for the best performance of the beam inside the capsule.

3.2 STRAWBERRY PROCESSING

The flowchart in Figure 6 shows the steps followed during the study.

When they arrived, they were packed in plastic trays and divided into portions of 250g each. When they arrived, they were packed in cardboard boxes, each

properly sanitise them, eliminating dust residue, soil and a large part of the microbiological load, as shown in Figure 7.

Figure 7. Washing the strawberries under running water.
Source: Prepared by the author, 2015.

3.2.2 Fruit sanitisation

The fruit was sanitised as shown in the flowchart. Treatment 1 (T1) is the control sample, which was only washed with running water. In test 2 (T2), the samples were sanitised with a 150 ppm sodium hypochlorite solution for 15 minutes. Test 3 (T3) used water with 100 ppm ozone (5 minutes). In test 4 (T4) the samples were treated with 200 ppm Ozone (10 minutes), and finally, in test 5 (T5) after washing the strawberries with running water, they were exposed in a laminar flow hood with a Philips TUV 15 W / G15 78 UV radiation lamp, with a peak at 253.7 nm, with germicidal action proven by the manufacturer (Figure 8). The parameters for using ozone were based on the capacity of the ozoniser.

Figure 8. Different strawberry sanitisation and UV radiation treatments.

Source: Prepared by the author, 2015.

Figure 8A shows the strawberries already cleaned and separated for sanitisation.

3.3.2   Total titratable acidity index   (ATT)

The determination of ATT was expressed as a percentage of citric acid and follows the determination of the Adolf Lutz Institute (2008), in which 10g of the fruit were macerated with a sieve and fork, forming a concentrated juice. Then 10 mL of strawberry juice was transferred to an Erleynmeyer flask with 90 mL of distilled water. 3 drops of Bromocresol green indicator were added and titration was carried out under agitation with 0.1N NaOH solution until the solution changed colour. The analyses were carried out in triplicate. The result was expressed as g% citric acid according to Equation 2.

A = 0.0064 x V x 100 / G   (Equation 2)

Where:

A = grams % citric acid

G = 10 mL of sample

V = volume of NaOH used.

3.3.   3pH

To determine pH, an AAKER digital bench pH meter was used, calibrated with buffer solutions 4 and 7 before the measurements began, and the results were obtained from direct readings of the juice pulp. The pH evaluations followed the determinations of the Adolf Lutz Institute (1985). The analyses were carried out in triplicate.

3.3.   4Total Soluble Solids (Brix)

Two drops of strawberry pulp were used to determine total soluble solids (TSS).

The TSS content was obtained by refractometry using a portable refractometer (ABBE Model WYA-2S) with a reading range between 0 and 100 °Brix. The results were expressed in °Brix, with temperature correction. The analyses were carried out in triplicate.

3.3.   5Mass Loss

For the mass loss tests, carried out after sanitising the strawberries and identifying the packaging, 3 samples were individually weighed from each of the 5 tests over the 10 days of storage, totalling 150 samples. The individual weight of each previously identified sample was noted down so that its weight could be deducted from the following days' weighings.

So, on the first day, the weight loss is zero, as there is nothing to compare it to, assuming that the mass loss starts to be counted from the second day of the tests.

With the aid of a GEHAKA BK600 analytical balance (0.001g resolution), the results were expressed as a percentage of mass loss according to the

Equation 3:

$$\% \, PM = [(\, Mi - Mf) \, / \, Mi] *100 \qquad \text{(Equation 3)}$$

Where:

% PM = percentage of accumulated partial mass loss.

Mi = initial mass of the sample at a given time in g.

Mf = final mass of the sample in the period following Mi in g.

3.3.   6Microbiological analysis

The analyses were carried out on 25 g samples of strawberries from each treatment, weighed into sterile plastic bags in an aseptic environment and homogenised with 225 mL of 0.1% peptone water in a stomacher (Marconi®). Successive dilutions were then made in 0.1% peptone water. Using 1mL per tube per dilution and 0.1mL of the tubes on plates with Potato Dextrose Agar (BDA) acidified with 10% tartaric acid to pH 3.5 with subsequent incubation at 25°C for 10 days in B.O.D LUCA-161/01, in accordance with Normative Instruction n°62 of 26 August 2003 (BRASIL, 2003).

Predominant fungal colonies on plates containing between 15 and 150 colonies were selected to identify contaminants. The results were expressed in logarithmic cycles (log CFU.g^{-1}).

3.4 STATISTICAL ANALYSIS

The results were statistically analysed using analysis of variance (ANOVA) in Microsoft Excel 2013, Action tool version 2.9.29.368.534, June 2015. Data is presented as mean ± standard deviation. Dunnett's test was used to estimate the minimum significant differences between the means of each test compared to the control sample at a 5% significance level.

CHAPTER 4

RESULTS AND DISCUSSIONS

4.1 EVALUATION OF THE OZONE GENERATOR'S FUNCTIONALITY

The device produces a noise characteristic of an electric shock, and as proof that the equipment has been correctly assembled, a bluish discharge will appear between the two wires that are attached inside the capsule. The characteristic smell of ozone was felt at the outlet of the hose. This really proved that the equipment was in ideal working order for the strawberry sanitisation experiment.

The device produces little ozone compared to commercial models, but if the outlet is placed to bubble in a solution of potassium iodide, within a few minutes a darkening can be seen due to the release of iodine, a typical reaction for identifying ozone (ARMAROLI, 2007).

In order to get the ozonator to work better, several tests were carried out by varying the potentiometer versus the voltage setting of the power supply. The best discharge of the corona effect was obtained with the potentiometer supplying a flow rate of 2.55 L.min^{-1} and the power supply with a voltage of 29.9 V.

According to Williams (2015). Atmospheric air is made up of 78.08% nitrogen, 20.95% oxygen, and 0.97% other gases with a minority share. It is therefore possible to calculate the theoretical amount of ozone formed after leaving the ozoniser.

2.55 L/min = 100%

x L/min=20.95%

Making the proportional calculation:

[((2.55 L/min*20.95%))/(100%)]=0.53 L/min

It is therefore possible to assume that 0.53 L.min^{-1} of oxygen passes into the capsule.

Figure 10 shows the corona effect inside the capsule through which the air circulates.

Figure 10. Corona effect inside the capsule through which the air circulates.
Source: Prepared by the author, 2015.

The thickness of the capsule is 2 cm high by 2.5 cm wide, multiplying this gives an area of 5 cm^2 in which the oxygen must pass perpendicularly; of this area, the corona effect "occupies" around 20%. So if 100 per cent of the oxygen passing through the area occupied by the electric discharge is converted into O3, we move on to calculating the O_3 flow rate.

0.53 L/min=100%

x L/min=20%

Calculating proportionally, we find that x is equivalent to 0.10 L.min^{-1} of O_2 which is converted into O_3. To calculate the concentration of ozone used to sanitise the strawberries, we need to multiply this amount by the density (2.14 g.L^{-1}) of the O_3, the contact time (5 minutes in T3 and 10 minutes in T4) with the fruit and the amount of water (10 litres) for each treatment. As calculated below:

Calculation for T3:

[Ozone]= [((0.10 L/min*2.14 g/L*5min))/(10 L)]

[Ozone]=0.107 g/L=107 ppm

Calculation for T4:

[Ozone]= [((0.10 L/min*2.14 g/L*10min))/(10 L)]
[Ozone]=0.214 g/L=214 ppm

This adjustment to the ozoniser's operating conditions resulted in a quantity of ozone generated of approximately 107 ppm for T3 and 214 ppm for T4 (Figure 11).

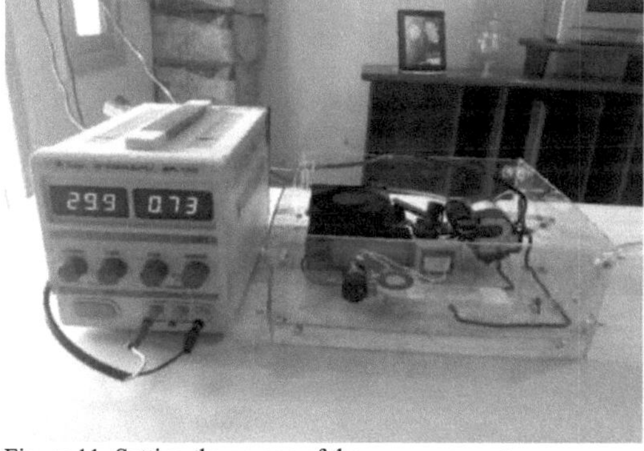

Figure 11. Setting the source of the ozone generator.
Source: Prepared by the author, 2015.

4.2 EVALUATION OF MASS LOSS

The mass loss control test always used the control treatment as a comparison parameter, in order to observe the behaviour of the sanitiser on the strawberry product without the presence of chemical agents.

Nadas, Olmo & García (2003) applied 1.5 ppm of gaseous ozone to strawberries

For treatment T4, graph (C), in the sample on day 9 the loss value reached 1.10%, falling to 0.81% on day 10. It is therefore difficult to predict what the correct behaviour would be, as the curve rises more sharply on the ninth day.

In graph (D), the values for T5 were very close to T1, slightly lower than the control treatment.

The calculation of the analysis of variance and Dunnett with a 95% confidence level, in which each treatment is compared individually against the control sample,

in all the comparisons it was observed that all the averages are statistically equal (p>0.05) as there was no significant difference.

Unlike Holtz (2006), who found that strawberries treated only with ozone showed significantly lower mass loss compared to the use of a 200 ppm chlorine solution and cassava starch and corn starch films during storage. In the treatments applied with the use of ozone, there was a slight increase compared to the use of chlorine.

According to Fennema et al. (2010) even losses as small as 1% in strawberries can result in significant losses in surface gloss. In the study in question, the best result for mass loss was observed in T5, which was 0.87%.

4.3 PH RESULTS

The pH monitoring during the storage period always used the control treatment as a parameter for comparison. Figure 13 shows the pH behaviour curves.

In graph (A), on all storage days, T2 obtained higher pH values than T1, and the pH values of the sanitised strawberries were no higher than 3.83 during storage at 10°C. Although a decrease in the pH values of the strawberries was observed from the 4th day of storage onwards, this behaviour was similar for all treatments.

In graph (B) T1 and T3 had very close values, with T3 having pH values slightly higher than T1 on the 10th day. In graphs (C) and (D) T4 and T5 alternated values in the first few days, but similarly to T3 ended up with slightly higher pH values.

Figure 13. Monitoring pH during storage.

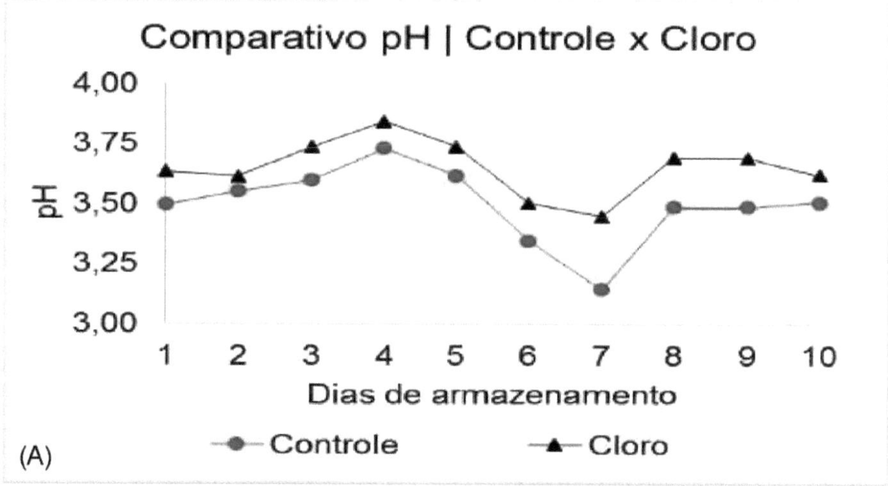

(A)

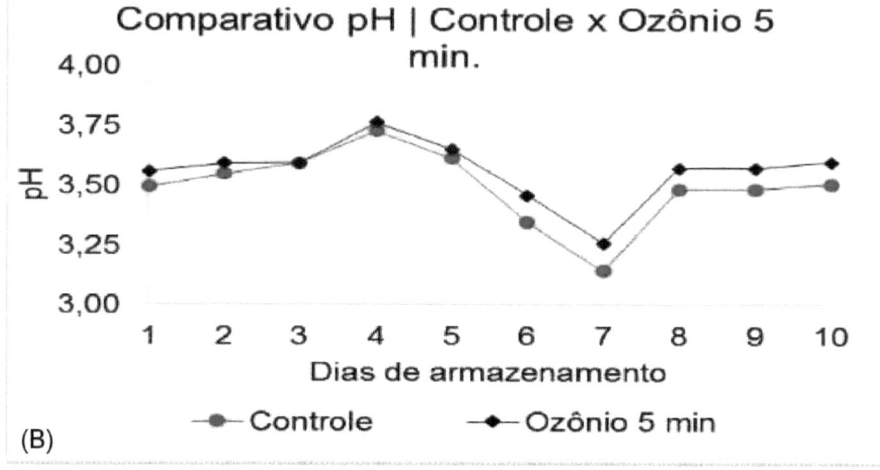

(B)

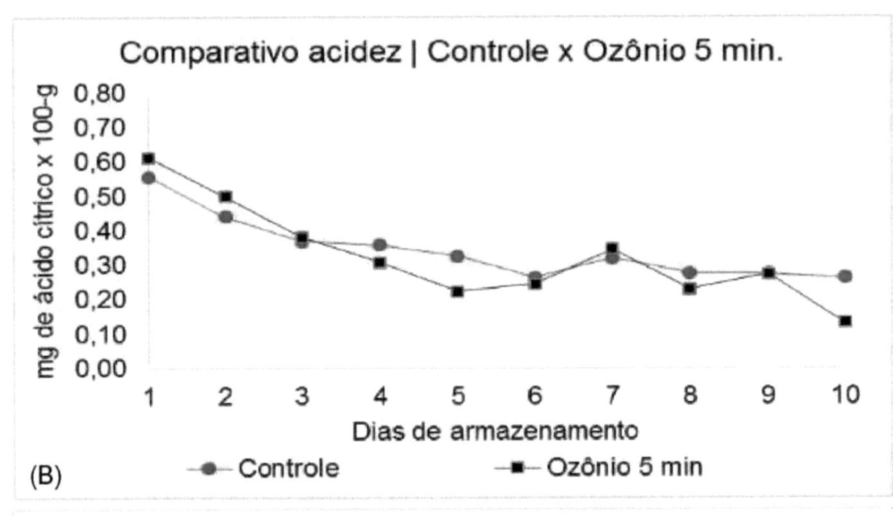

Comparativo acidez | Controle x Ozônio 5 min.
mg de ácido cítrico x 100-g
0,80
0,70
0,60
0,50
0,40
0,30
0,20
0,10
0,00
1 2 3 4 5 6 7 8 9 10
Dias de armazenamento
Controle
Ozônio 5 min
(B)

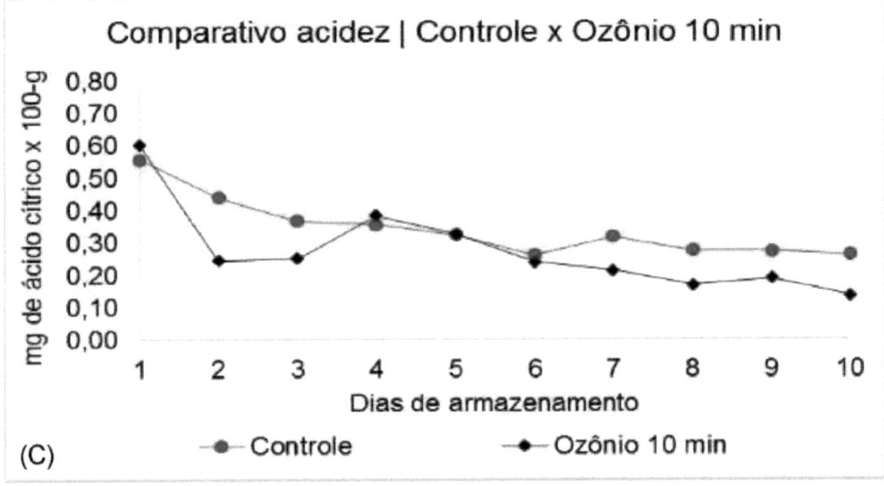

Comparativo acidez | Controle x Ozônio 10 min
mg de ácido cítrico x 100-g
0,80
0,70
0,60
0,50
0,40
0,30
0,20
0,10
0,00
1 2 3 4 5 6 7 8 9 10
Dias de armazenamento
Controle
Ozônio 10 min
(C)

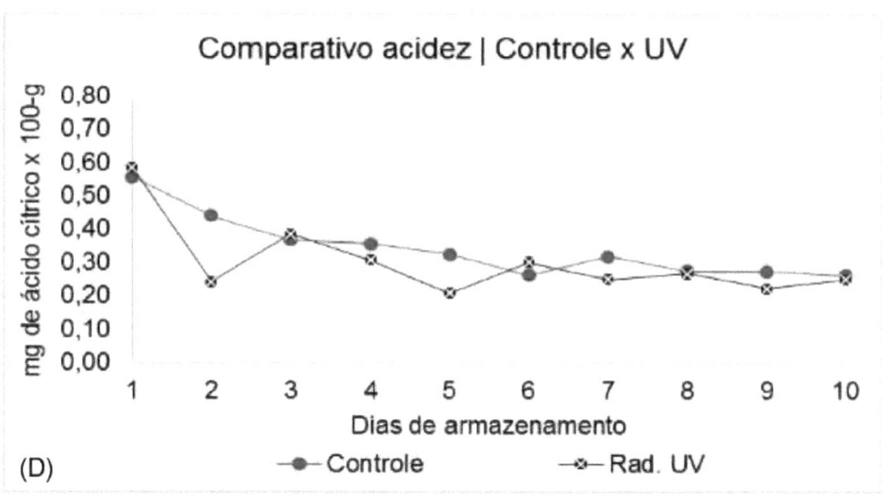

Source: Prepared by the author, 2015.

The reduction in acidity is due to the degradation of the organic acids present in the strawberry, which can lead to a loss of nutrition and vitamins in the food. The total titratable acidity values (mg.100g^{-1} of pulp) obtained, between 0.13 and 0.61, are lower than those described in the literature by Domingues (2000) who found between 0.85 and 0.99 mg.100g^{-1}. However, the differences observed may be due to factors such as the type of cultivar and the stage of ripeness of the fruit.

Soares et al. (2011) evaluated the ATT at two stages of ripeness of the strawberry variety 'Camino Real' with red and ripe % and obtained values of 0.760 and 0.867 mg of citric acid per 100 g of pulp.

The Dunnett test values showed no significant differences for the different treatments at a 95% confidence level.

4.5   RESULTS ON TOTAL SOLUBLE SOLIDS (TSS)

According to Chitarra (2005), strawberries have around 7% total soluble solids. Figure 15 shows the behaviour of this quality index over the storage period.

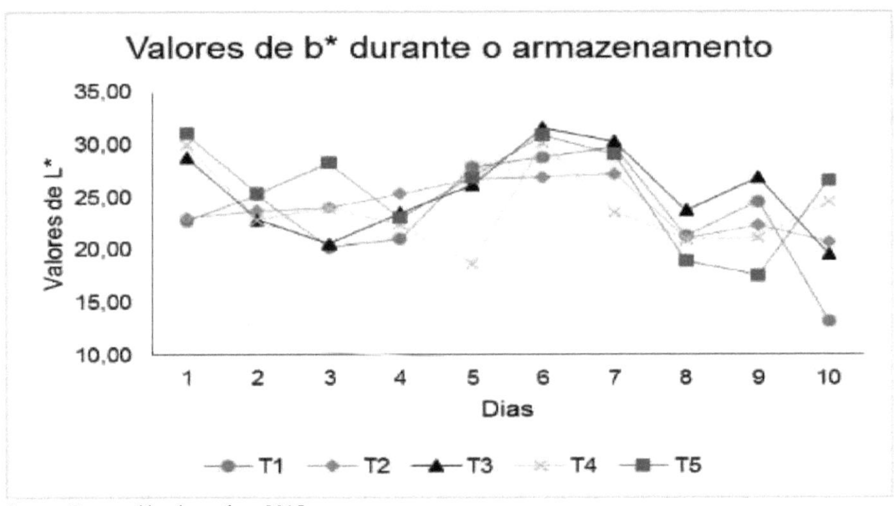

Source: Prepared by the author, 2015.

The Dunnett test values showed no significant differences between the different treatments and the control sample at a 95% confidence level.

In general, there was an increase in L* and a* values from the 5th day of storage, while there was a more linear variation in b*, with T2 being the most stable treatment over time. However, the colour intensity remained below that of treatments T3 and T5.

Soares et al. (2011) observed the L* parameter of the skin between 33.02 and 25.93; a* of the skin between 31.35 and 27.81 and b* of the skin between 24.21 and 14.55, similar to T1 on the tenth day of storage.

Monitoring the colour difference during storage becomes an important practice, due to its linear relationship with the formation of dark pigments, and is important for assessing the colour changes caused by the browning reaction. A total increase in colour difference means that the product has lost the characteristic colour of the initial sample (YAMASHITA et al., 2006).

Table 4: L*a*b* values converted into RGB for visual analysis.

Comparison of mass loss between sanitisation tests					
Storage days		Type of sanitiser			
	Control	Chlorine	Ozone 5 mins	Ozone 10 mins	Rad. UV

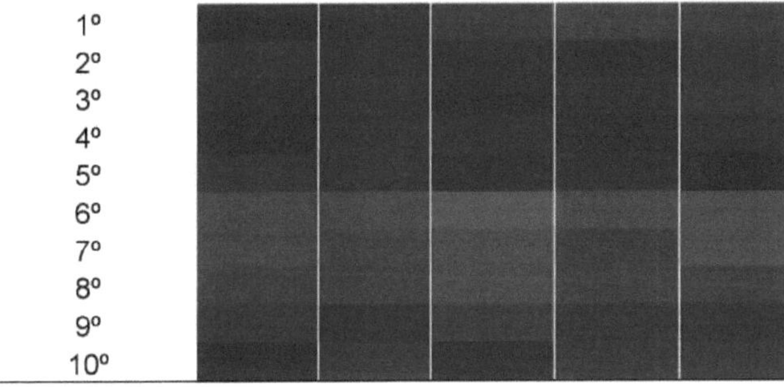

Source: Prepared by the author, 2015.

As shown in Figure 16, Table 4 confirms the change in colour from day 5 onwards, with a lighter colour tone, returning to darker tones on day 8. Table 4 also shows that the variations in colour and RGB colour intensity were greater as the storage time increased. The colour intensity was higher for T5 and lighter for T3. The results can be visualised qualitatively in Figure 17.

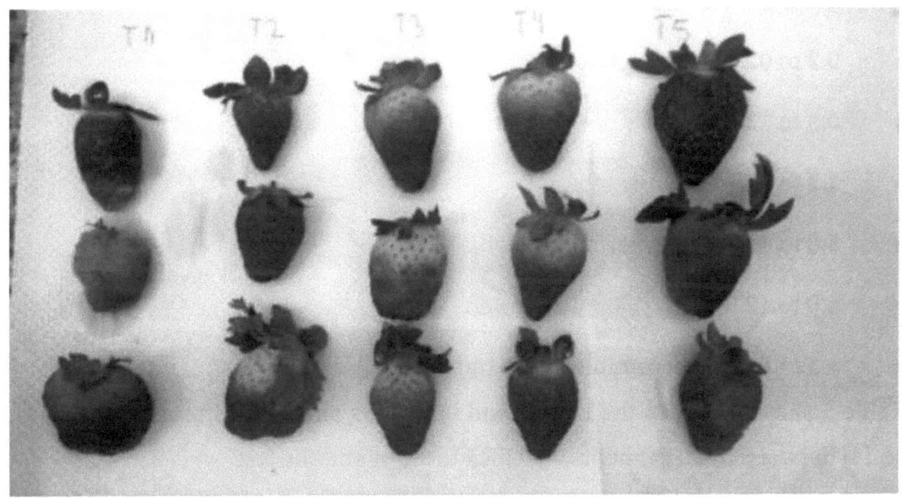

Figure 17. Assessment of fruit colour at the end of storage.
Source: Prepared by the author, 2015.

On the last day of storage (day 10 of the test), the colour of the strawberries was checked visually. T1 already had mould growth on the surface and a dark colour. T5 had good flesh firmness but dry patches. Visually, the best results were between T2, T3 and T4.

Table 5 below shows the colour difference using the AE* calculation, always

Control	4,75	4,10	4,25	4,65	4,75	5,65	7,45	7,55	8,00	8,00
Chlorine	2,10	5,10	5,70	4,55	4,55	5,05	4,00	5,00	7,00	7,15
Ozone 5 min	2,80	3,05	3,30	4,30	4,50	3,00	4,00	5,85	5,05	5,00
Ozone 10 min	2,70	3,50	4,45	4,50	3,00	4,80	4,00	6,55	5,05	5,00
Rad. UV	3,75	3,30	3,20	5,60	5,55	4,85	6,30	6,55	8,00	8,00

Source: Prepared by the author, 2015.

The reduction in the initial count of filamentous fungi and yeasts obtained in this study with aqueous ozonation of strawberries at 107 ppm and 214 ppm was satisfactory, given that the gas reacts with the components of the strawberries, limiting its action on the contaminating micro-organisms. Margosan (1998) managed to inhibit the germination of spores of B. cinerea, M. fructicola, P. digitatum and R. stolonifer using aqueous ozone with a solution at a concentration of 0.13 ppm for 80 minutes, directly on these fungi, with no other components interfering with the action of the sanitiser.

The effect of UV on bacteria and fungi such as Aspergillus and Penicillium was reported by Kleczkowski (1968) for the preservation of papaya. UV irradiation exhibits an effect on disease resistance and post-harvest disease control of crops.

Chlorine, in turn, can react with organic residues and result in the formation of potentially mutagenic or carcinogenic products (CHANG et al., 1988). In addition, ozone can be produced at the place where it will be used, reducing transport and storage costs for sanitisers (KHADRE et al., 2001).

CHAPTER 5

CONCLUSION

The experience of researching, designing and assembling an ozone generator that had the expected effect was of great importance, making it possible to review basic engineering concepts such as the operation of electrical equipment.

Although there were no statistically significant differences between any of the treatments, the physicochemical tests evaluated were essential for assessing the quality of the strawberries during storage, so that even with the use of chemical agents for biological control, the fruit retained the characteristics of the control sample. This does not de-characterise them, which means that even after minimal processing, they have similar amounts of nutrients to products that have not undergone any type of processing.

The microbiological analyses showed that the best agents for controlling mould and yeast were aqueous ozone, used for 5 minutes at 107 ppm and 10 minutes at 214 ppm. The results were satisfactory compared to the use of chlorine and ultraviolet radiation. However, the excess water on the surface of the fruit before packaging may have had a negative influence on the counts, as it acts as an aid in the development of the biological load, increasing the free water activity on the surface of the fruit. As a future proposal, along with the water draining stage after sanitisation, the fruit could be fumigated with only ozone gas circulating over them, so as well as further reducing the microbiological load, it would be possible to reduce the surface free water, which would not favour fungal development.

CHITARRA, M. I. F. Post-harvest of fruit and vegetables: physiology and handling. 2ª Revised and expanded edition. Lavras, MG - UFLA. 2005.

COELHO, N. R. A. Notions of hygiene in the food industry, Fruit and Vegetable Processing - Catholic University of Goiás. Food Engineering Course. 2014. Available at < http://wp.ufpel.edu.br/mlaura/files/2014/02/Higieneiza%C3%A7%C3%A3o-na-ind%C3%BAstria-de-alimentos.pdf> accessed on 13/09/15.

CORTEZ, L. A. B.; HONÓRIO, S. L.; MORETTI, C. L. Cooling fruit and vegetables: Embrapa Hortaliças - Brasília, DF. 60 - 64 p. 2002

DOMINGUES, D. M. Effect of gamma radiation and packaging on the conservation of "Toyonoka" strawberries stored under refrigeration. Piracicaba, 2000. 58p. Dissertation - (Master's Degree in Nuclear Energy in Agriculture), Escola Superior de Agricultura "Luiz de Queiroz", Universidade de São Paulo.
EMBRAPA - Brazilian Agricultural Research Corporation; Post-Harvest Pathology, Tropical Fruits, Vegetables and Ornamentals. Brasília - DF. 2006.

FDA - Food and Drug Administration; Guidance for Industry: Guide to Minimise Microbial Food Safety Hazards of Fresh-cut Fruits and Vegetables. Centre for Food Safety and Applied Nutrition, February 2008. Available at:< http://www.fda.gov/Food/GuidanceRegulation/GuidanceDocumentsRegulatoryInformation/ucm064458.htm> Accessed on: 28/08/2015

FDA - Food and Drug Administration; Food Additive Status List. Centre for Food Safety and Applied Nutrition, December 2014. Available at:

http://www.fda.gov/food/ingredientspackaginglabeling/foodadditivesingredie nt s/ucm091048.htm. Accessed on: 28/08/2015.

FENNEMA, O. R.; DAMODARAM, S.; PARKIN, K. L. Fennema's Food Chemistry. 4 ed. - Porto Alegre: Artmed, 2010.

FRANÇOS, I. L. T.; COUTO, M. A. L.; CANNIATTI-BRAZACA, S. G. Physical-chemical alterations in irradiated and stored strawberries (Fragaria anassa Duch.) Campinas, SP. jul.-set. 2008.

FREITAS-SILVA, O.; VENÂNCIO, A. Ozone applications to prevent and degrade mycotoxins: A review. Drug Metabolism Reviews, v.42, p.612-620, 2010.
GERMANO, P. M. L. Food hygiene and health surveillance: quality of raw materials, food-borne diseases, training of human resources. - 3ª revised and expanded edition. - Barueri, SP: Manole, 2008.

GIAMPIERI, F.; TULIPANI, S.; ALVAREZ, J.; QUILES, J.; MEZZETTI, B.; BATTINO, M.; The strawberry: Composition, nutritional quality, and impact on human health. Nutrition, 2012. p. 9-19

GONSALVES, P. E. Food book. Editora Martins Fontes - São Paulo, SP. 1992.

GUZEL-SEYDIM, Z. B.; GREENE, A. K.; SEYDIM, A. C. Use of ozone in the food industry. Lebensmittel Wissenschaft und Technologie, San Diego, v. 37, n. 4, p. 453-460, 2004.

and hydroponics. Vitória: PROSAB, 2003. p. 169-208.

LEME, F. P. Teorias e técnicas de tratamento de água. Rio de Janeiro: ABES, 1990.
LOVATEL, J. L.; CONSTANSI, A. R.; CAPELLI, R. Fruit and vegetable processing.
Caxias do Sul, RS: Educs, 2004. p 35.

LIU, J.; STEVENS. C.; KHAN, V. A.; LU, J. Y.; WILSON, C. L.; ADEYEYE, O.; KABWE, M.
K.; PUSEY, P. L.; CHALUTZ, E.; SULTANA, T.; DROBY, S. Journal of Food Protection.
Application of ultraviolet-C light on storage rots and repening of tomatoes.
Desmoines, Vol 56, n. 10, 1993. 868-872 p.

MADAIL, J. C. M.; MIGLIORINI, L. C.; REICHERT, L. J.; Sistemas de Produção,
Conservação de morango para a elaboração de produtos industrializados.
Embrapa clima temperado. Electronic version - November 2005.

MARGOSAN, D. A. and SMILANICK, J. L. Mortality of spores of Botrytis cinerea,
Monilinia fructicola, Penicillium digitatum, and Rhyzopus stolonifer after
exposure to ozone under humid conditions. Phytopathology, v. 88, p. 858, 1998.

MORETTI, Celso Luiz. Manual of minimum processing of fruit and vegetables.
Brasília: Embrapa Hortaliças and SEBRAE, 2007.

NADAS, A.; OLMO, M.; GARCÍA, J. M. Growth of Botrytis cinerea and strawberry
quality in ozone-enriched atmospheres. Journal of Food Science, v. 68, n°. 5, p.
1798-1802, 2003.
OECD-FAO. Agricultural Perspectives in Brazil: challenges for Brazilian

agriculture 2015-2024. Chapter 2. Brazilian Agriculture: Prospects and Challenges 2015.

OZONE SOLUTIONS, Ozone Use on Fruits and Vegetables, July 2014. Available at: <http://www.ozonesolutions.com/info/ozone-fruits-vegetables> Accessed on 20/09/15.

PEREIRA, A. M. S.; LUCA, S. J. Analysis of trihalomethanes in public water supply. BRAZILIAN CONGRESS OF ENVIRONMENTAL SANITARY ENGINEERING, 15. Belém. Proceedings. Rio de Janeiro: ABES, 1989.

PERERA, C. O. Minimal Processing of Fruits and Vegetables, In: Handbook of Food Preservation. Second edition, 2007.

PINHEIRO, N. M. S.; FIGUEIREDO, E. A. T.; FIGUEIREDO, R. W.; MAIA, G. A.; SOUZA, P. H. M. Evaluation of the microbiological quality of minimally processed fruit sold in supermarkets in Fortaleza. Revista Brasileira de Fruticultura. 2005.

RAHMAN, M. S. Handbook of food preservation. 2nd ed. CRC Press, 6000 Broken Sound Parkway NW, Suite 300. 2007. p 287.

RAHMAN, M. S; KLEZKOWSKI, A. Methods of inactivation by ultraviolet radiation, Methods in Virol. 1968. In:. Handbook of food preservation. 2nd ed. CRC Press, 6000 Broken Sound Parkway NW, Suite 300, 2007.
REIS, K. C.; SIQUEIRA, H. H.; ALVES, A. P.; SILVA, J. D.; LIMA, L. C. O. Effect of different sanitisers on the quality of strawberry cv. Oso grande. Ciênc. agrotec.

CHAPTER 7

APPENDIX A - RESULTS OF MASS LOSS ANALYSES

Percentage of mass loss during storage time

Storage days	Means of triplicates ± Standard Deviation				
	Type of sanitiser				
	Control	Chlorine	Ozone 5 min	Ozone 10 min	Rad. UV
2°	0,1224 ± 0,0435	0,0345 ± 0,0104	0,0598 ± 0,0107	0,0614 ± 0,0059	0,0568 ± 0,0095
3°	0,1695 ± 0,0791	0,1790 ± 0,0397	0,1316 ± 0,0247	0,1397 ± 0,0727	0,0912 ± 0,0200
4°	0,4131 ± 0,0708	0,1475 ± 0,0603	0,1821 ± 0,0515	0,1803 ± 0,0693	0,1839 ± 0,0943
5°	0,2146 ± 0,0813	0,1716 ± 0,0824	0,2257 ± 0,0751	0,1584 ± 0,1282	0,1780 ± 0,0610
6°	0,5584 ± 0,2372	0,5849 ± 0,2373	0,4123 ± 0,0696	0,3512 ± 0,0471	0,3397 ± 0,0215
7°	0,4221 ± 0,0501	0,3005 ± 0,1099	0,4033 ± 0,1029	0,5082 ± 0,1697	0,4463 ± 0,1069
8°	0,4187 ± 0,1210	0,3486 ± 0,0897	0,5033 ± 0,0534	0,3870 ± 0,0621	0,4888 ± 0,0664
9°	0,6938 ± 0,0125	0,9165 ± 0,1429	0,7464 ± 0,1199	1,1036 ± 0,4465	0,7228 ± 0,0830
10°	1,0066 ± 0,3807	1,0947 ± 0,1966	1,3263 ± 0,1712	0,8100 ± 0,1489	0,8723 ± 0,1707

Dunnett's test for loss of mass between treatments and control sample

Difference between Levels	Average	P-Value
T2 - T1	-0,026836	0,99970
T3 - T1	-0,003162	0,99999
T4 - T1	-0,035479	0,99915
T5 - T1	0,090888	0,97043

Confidence intervals in relation to the control sample for mass loss analysis

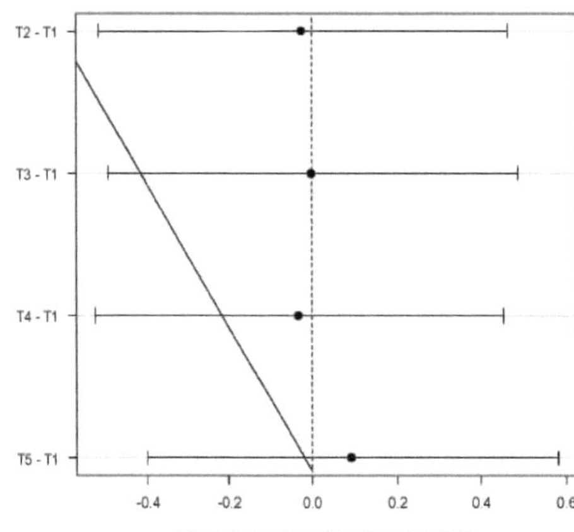

APPENDIX B - PH ANALYSIS RESULTS

Comparison of pH between sanitisation tests

Storage days	Means of triplicates ± Standard Deviation				
			Type of sanitiser		
	Control	Chlorine	Ozone 107 PPm	Ozone 214 PPm	Rad. UV
1°	3,49 ± 0,15	3,63 ± 0,01	3,56 ± 0,01	3,43 ± 0,00	3,54 ± 0,02
2°	3,55 ± 0,00	3,61 ± 0,03	3,59 ± 0,01	3,64 ± 0,00	3,75 ± 0,00
3°	3,59 ± 0,01	3,73 ± 0,07	3,59 ± 0,02	3,70 ± 0,08	3,58 ± 0,04
4°	3,73 ± 0,01	3,84 ± 0,02	3,76 ± 0,00	3,55 ± 0,03	3,67 ± 0,01
5°	3,61 ± 0,02	3,73 ± 0,02	3,65 ± 0,00	3,48 ± 0,02	3,53 ± 0,03
6°	3,34 ± 0,02	3,50 ± 0,02	3,46 ± 0,07	3,49 ± 0,01	3,45 ± 0,03
7°	3,14 ± 0,12	3,44 ± 0,02	3,26 ± 0,08	3,16 ± 0,12	3,38 ± 0,02
8°	3,48 ± 0,04	3,69 ± 0,02	3,57 ± 0,01	3,52 ± 0,02	3,60 ± 0,02
9°	3,48 ± 0,25	3,69 ± 0,01	3,57 ± 0,00	3,52 ± 0,01	3,60 ± 0,00
10°	3,50 ± 0,02	3,62 ± 0,01	3,60 ± 0,02	3,63 ± 0,00	3,60 ± 0,02

Dunnett's test for pH analysis in relation to the control sample

Difference between Levels	Average	P-Value
T2 - T1	0,155000	0,043488
T3 - T1	0,067667	0,621904
T4 - T1	0,020333	0,991154
T5 - T1	0,076667	0,517969

Confidence intervals in relation to the control sample for pH analysis

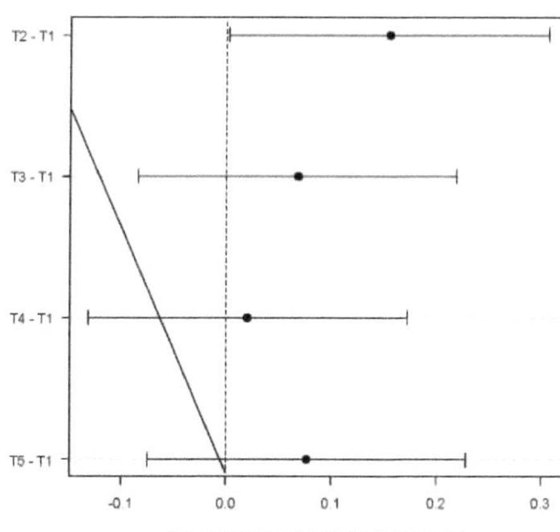

APPENDIX E - RESULTS OF THE COLORIMETRY ANALYSES

Dunnett's test for colourimetry analysis of L* values compared to control sample

Difference between Levels	Average	P-Value
T2 - T1	0,346333	0,999425
T3 - T1	1,375	0,907822
T4 - T1	0,238	0,999869
T5 - T1	1,287667	0,92518

Confidence intervals in relation to the control sample for L* colourimetry analysis

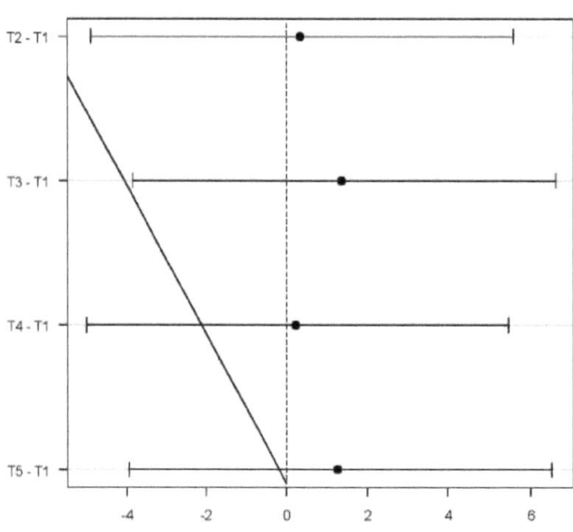

Dunnett's test for colourimetry analysis of a* values compared to the control sample

Difference between Levels	Average	P-Value
T2 - T1	2,555	0,647687
T3 - T1	2,102667	0,779013
T4 - T1	1,12	0,969016
T5 - T1	1,996333	0,807389

Confidence intervals in relation to the control sample for a* colourimetry analysis

I want morebooks!

Buy your books fast and straightforward online - at one of world's fastest growing online book stores! Environmentally sound due to Print-on-Demand technologies.

Buy your books online at
www.morebooks.shop

Kaufen Sie Ihre Bücher schnell und unkompliziert online – auf einer der am schnellsten wachsenden Buchhandelsplattformen weltweit! Dank Print-On-Demand umwelt- und ressourcenschonend produziert.

Bücher schneller online kaufen
www.morebooks.shop

Printed by Books on Demand GmbH, Norderstedt / Germany